CHAINS, WEBS & PYRAMIDS
the flow of energy in nature

CHAINS, WEBS

WRITTEN BY LAURENCE PRINGLE

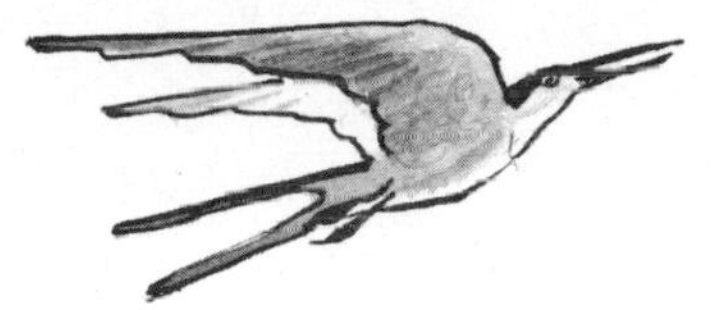

& PYRAMIDS

the flow of energy in nature

ILLUSTRATED BY JAN ADKINS

Thomas Y. Crowell Company, New York

The author and the publisher are grateful to the Macmillan Company for permission to draw upon material which appeared in different form in Mr. Pringle's *Ecology Science of Survival,* published by Macmillan in 1971.

 Published simultaneously in Canada by Fitzhenry & Whiteside Limited, Toronto.
Designed by Jan Adkins
Manufactured in the United States of America

Library of Congress Cataloging in Publication Data

Pringle, Laurence P. Chains, webs & pyramids. SUMMARY: Describes the steps in a food chain and discusses their importance in the maintenance of life. 1. Food chains (Ecology)—Juv. lit. [1. Food chains (Ecology) 2. Ecology] I. Adkins, Jan, ill. II. Title. QH541.14.P74 574.5′3 75-1084
ISBN 0-690-00562-8
0-690-00563-6(LB)

1 2 3 4 5 6 7 8 9 10

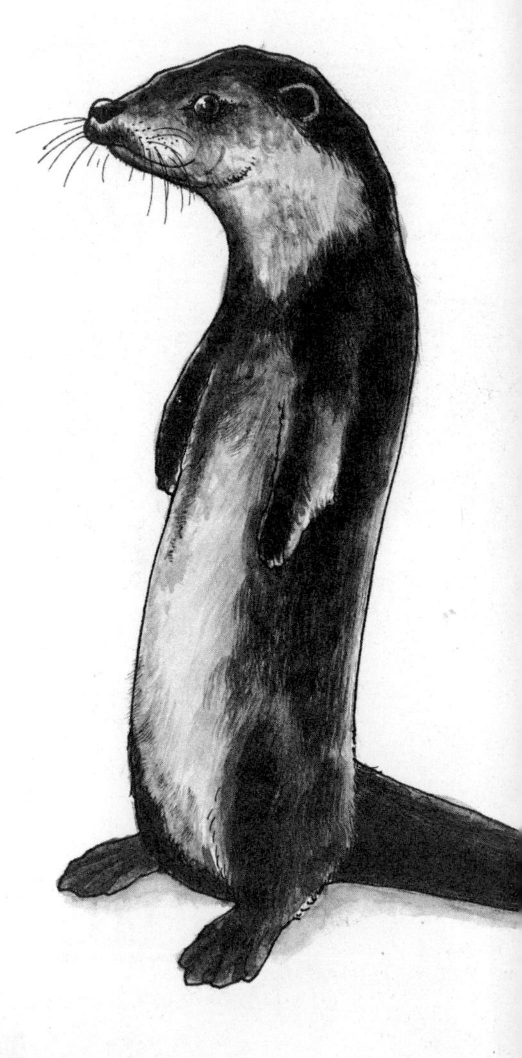

CHAINS, WEBS & PYRAMIDS
the flow of energy in nature

IT WAS DUSK ALONG THE RIVER. Night animals began to prowl. A raccoon waded into shallow water and turned over a stone with its paw. A crayfish darted out. The raccoon grabbed it, bit off its head, then ate the rest. As the raccoon fed, it took in energy that was stored in the body of the crayfish.

All living things need energy in order to stay alive. An animal needs energy as it breathes and moves. A plant uses energy as it grows, as the petals of a flower open, and as seeds develop. All living things need energy, and all living things contain energy.

Every second, all over the earth, energy is flowing from some living things to other living things. Sometimes this flow of energy is part of an exciting event, as when a raccoon catches a crayfish. Usually, however, the flow of energy goes unnoticed or is taken for granted. Energy flows when a bit of leaf is eaten by an earthworm, when a cow grazes on grass, and when you eat a sandwich.

Whenever an animal eats a plant, or eats another animal, it receives some food energy. It also gets some nutrients. Nutrients are minerals and other elements that are needed by plants and animals for normal growth and development. They are part of all living things. When a plant or animal dies and decays, most of its nutrients are released into the soil, water, or air. Often the nutrients are taken up by another plant and begin another journey through living things.

These journeys are called cycles. Nutrients keep moving in cycles, from living things to nonliving things and back to living things, on and on. Your body may contain nutrients that were once part of a dinosaur, a butterfly, or a redwood tree. The nutrients of life are never destroyed or used up, but continually passed on.

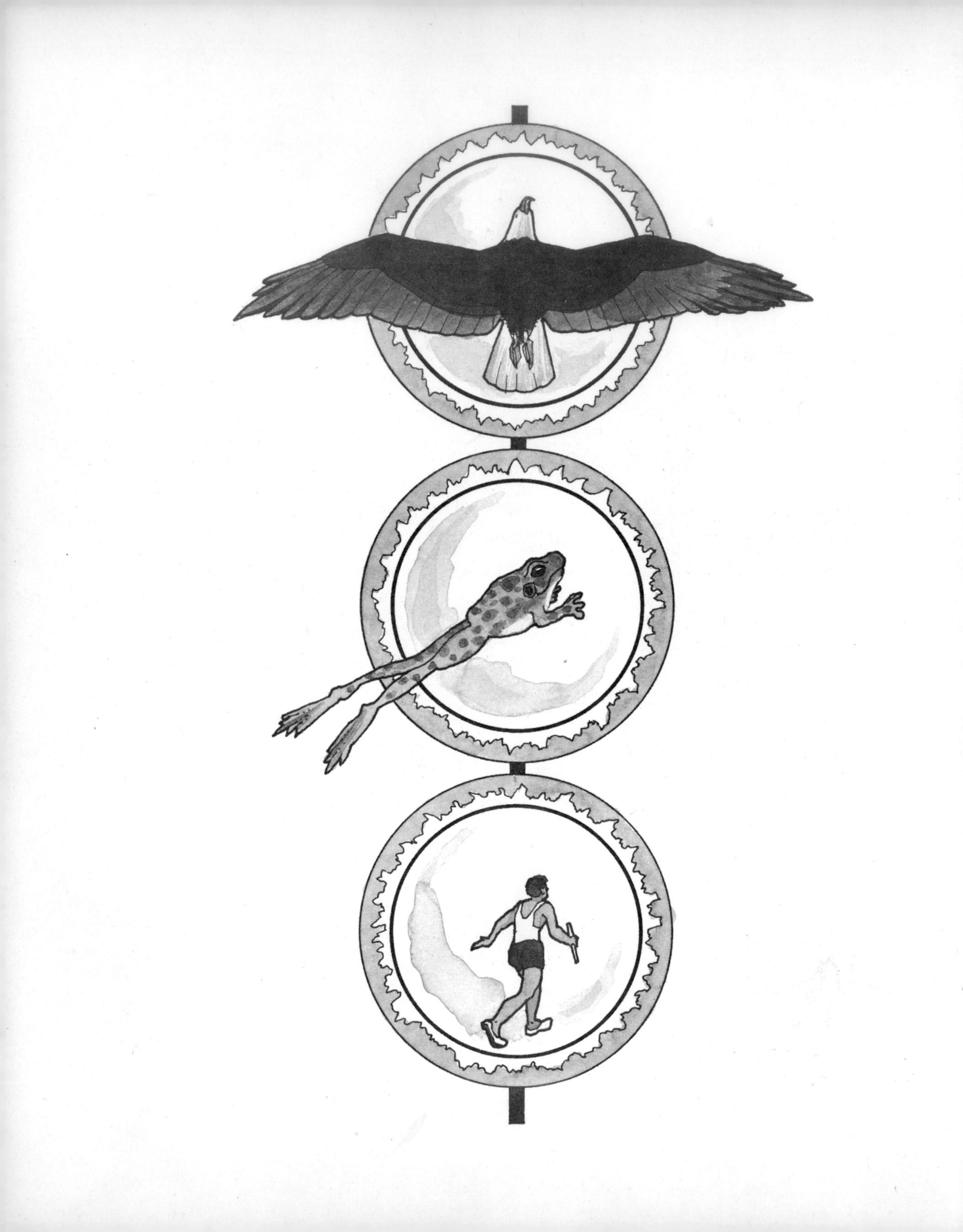

The *energy* of life is different. As it flows from one living thing to others, more and more of it *seems* to disappear. Actually it only changes into a form that is no longer useful for maintaining life. So more energy is needed. Plants and animals can live only because new supplies of energy constantly come from the sun. To trace the pathways of life's energy, you must begin at the sun.

Our sun is ninety-three million miles away. It gives off great amounts of energy. But only a small part of this energy reaches the earth. Near the earth, 50 percent of the sun's energy is reflected by clouds and dust back into space. Another 20 percent is absorbed by the atmosphere. About 30 percent heats the land and water. Some of it strikes the leaves of green plants. Of all the sunlight that shines on the earth, only about 2 percent is used by the leaves of green plants. Within these leaves something extraordinary happens. The sun energy is changed to food energy.

This food energy is the fuel that supports all life on earth. The sun powers the hop of a toad, the flight of an eagle, and the growth of a blade of grass—or of a person.

The process of changing sun energy to food energy is called photosynthesis, which means "putting together with light." That is what happens within green plants—a gas called carbon dioxide and water are "put together" with sun energy, forming simple sugars such as glucose. Later these simple sugars change to other forms of food energy within the plants. Now the energy is in a form that can be used by animals, including people.

Only green plants are able to trap the sun's energy and change it into food energy. Some of this energy is used by the plants as they grow and produce seeds. The rest is stored in the plant, and can be used by other living things.

When an animal eats a plant, it receives some of the plant's stored energy. That animal may be eaten by another animal. Energy flows from one living thing to another in a series of steps called a food chain.

Since no animal can make its own food, all food chains begin with green plants, the food producers. Some food chains have only two steps or links, such as corn→human. Whenever you eat corn, a cookie (made from wheat flour), an apple, or any other food that came from a green plant, you are part of a simple food chain.

Some food chains have several links: grass→mouse→snake→hawk. Food chains sometimes have as many as

five links, but most have only three or four. There is a good reason for this. As energy flows through a food chain it seems to disappear. Actually, energy can be changed but not destroyed or created. However, every time energy changes, some of it becomes heat energy which is given off into the air or water. So far as living things are concerned, this energy is useless.

A great deal of energy is lost in this way at each link in a food chain. Whenever an animal eats and digests part of a plant or animal, some of the energy it receives is lost as heat. Animals use large amounts of energy for moving about. Even while you are eating, you use energy to chew, swallow, and digest your food. As you sleep, energy is needed to keep your heart, lungs, and other organs working. And whenever stored energy is changed to working energy, some energy is lost as heat.

sun energy converted
to food energy by

ALGAE

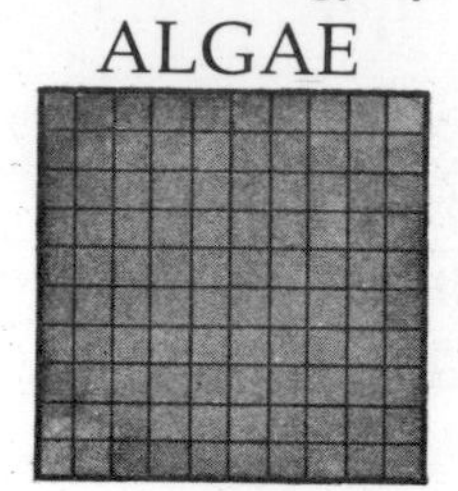

Since energy is lost at each link in a food chain, each organism passes on less energy than it receives. When a mouse eats some grass, it gets much less energy than the grass received from the sun. When a snake eats the mouse, it gets much less than the mouse received from the grass. The longer a food chain, the less energy there is available for the living things at the end of the chain. This limits the number of links in a food chain to four or five.

Scientists have studied real food chains in laboratories and outdoors. One scientist kept green algae plants and two kinds of small animals, hydras and water fleas, in aquariums. He was able to measure the amount of energy

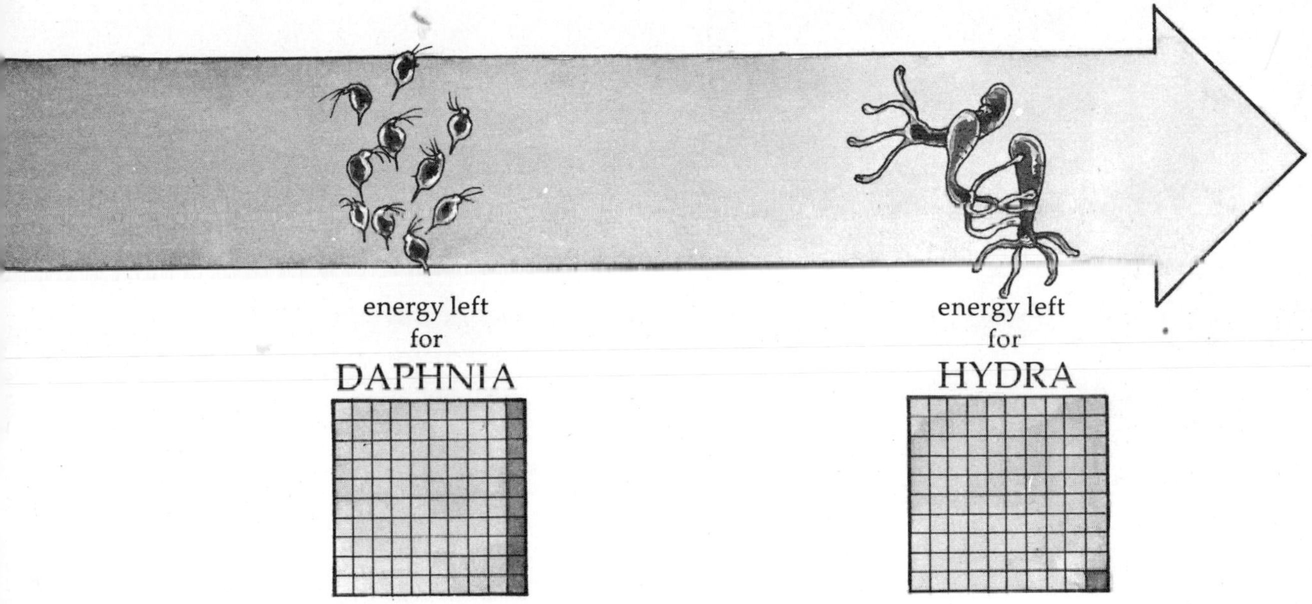

energy left
for
DAPHNIA

energy left
for
HYDRA

the algae plants produced by photosynthesis. When water fleas, called *Daphnia,* ate the algae, they received about 10 percent of the plant energy. And when the water fleas were eaten by hydras, the hydras received only 10 percent of the energy from them, or just 1 percent of the energy produced by the plants.

Recently scientists have been able to trace actual food chains in natural outdoor communities of plants and animals. They do this by using radioactive elements. These elements give off particles which cannot be seen or felt but which can be detected with special instruments. (The particles can harm living things, but are safe to use in very small amounts.)

In Tennessee, scientists put a tiny amount of radioactive phosphorus on the leaves and stems of some grass plants. Then for several weeks the scientists caught small animals that lived in the grass. They used a device to measure the amount of radioactive phosphorus within each animal. The scientists could tell when energy from the plants was within the animals because the radioactivity reached a peak then. Ants and small crickets reached a peak in radioactivity first. These insects were the most active plant eaters. Other plant-eating insects, such as grasshoppers, reached a peak after three weeks had passed.

Within a few days after the phosphorus was put on the plants, the scientists began to detect radioactivity in spiders. The spiders were predators. They killed and ate the plant-eating insects, and received energy, nutrients, and radioactive phosphorus from them. The radioactivity in spiders reached a peak after four weeks.

Other animals living in the grass community reached their peak of radioactivity about the same time as the spiders. Snails and beetles got food energy, nutrients, and phosphorus as they fed on dead grass leaves and perhaps on dead insects. Even though they ate only dead things, the energy they received could be traced back through food chains to its source—living green plants.

Food chains are a handy way to show how life's energy flows from one living thing to another. However, they show only a small part of the total energy flow in nature. Think of such a common food chain as grass→mouse→fox, for example. Mice are just one of dozens of different animals that eat grass. Foxes are just one of many animals that eat mice. Besides, there are other animals, such as fleas and ticks, which live *on* mice and foxes. Still others live *inside* mice and foxes. They all get a share of the flowing energy.

A food chain is incomplete in other ways. Whenever a living thing dies, it becomes food for bacteria and fungi. These simple plants cannot make their own food. They get food energy and nutrients by breaking down the remains of once-living plants and animals. Since bacteria and fungi cause these remains to decay, they are called decomposers. An accurate picture of energy flow in nature would show that decomposers get energy from every link in a food chain, once the organisms die.

If you could draw a picture of all the energy pathways in a field, it would be much more complicated than a few food chains. Hundreds of food chains would be joined and crisscrossed. You would have a food web.

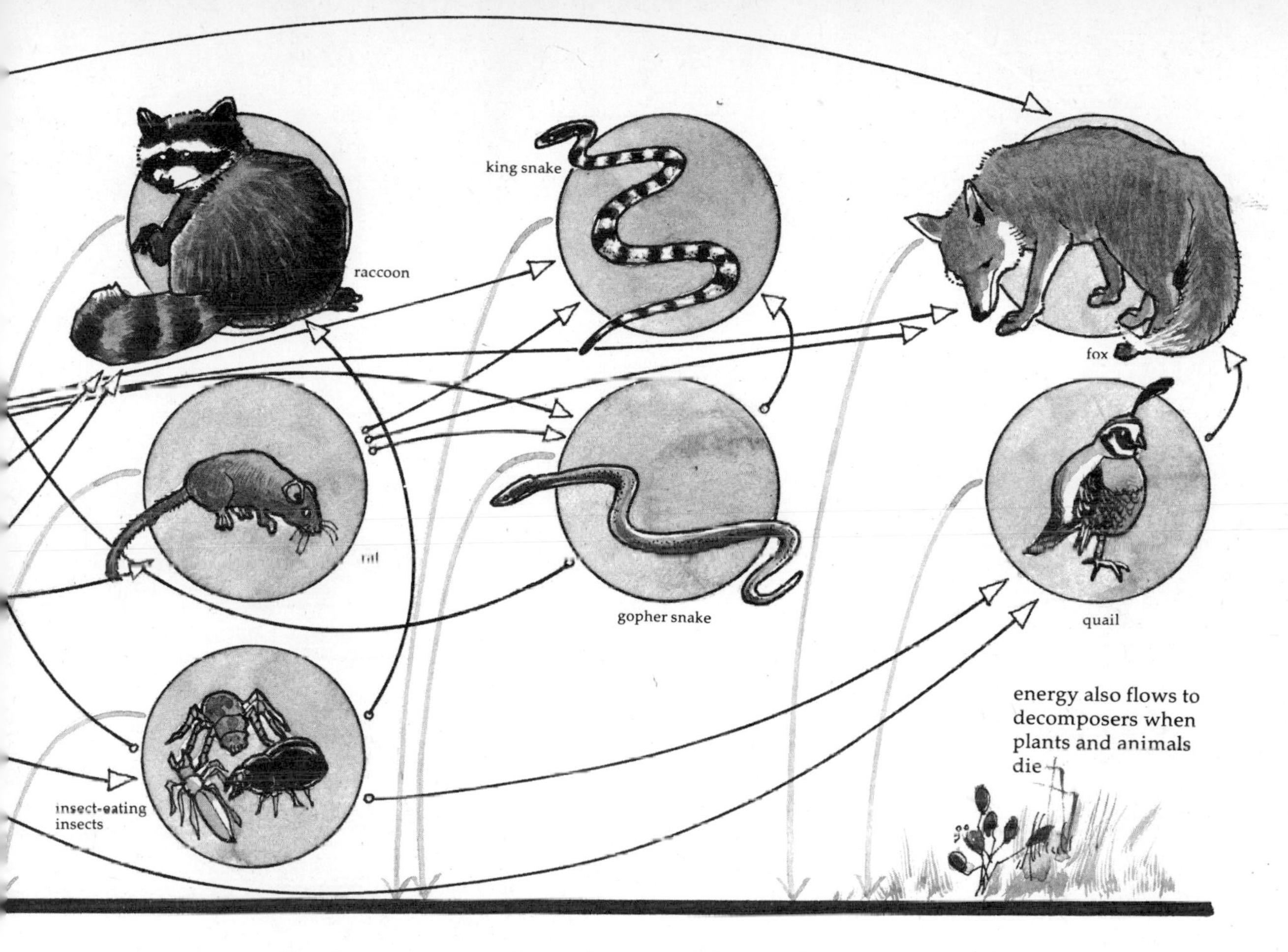

So far, no one has been able to figure out all the energy pathways in a real food web. In order to do that, scientists would have to know exactly how hundreds of different kinds of animals living in a field, pond, or other community received their food energy. Even a food web that shows only the flow of energy among a dozen animals and plants is quite complicated.

The drawings on these two pages show one such web: a simple food web in the dry chaparral forests of California. The drawing on the next two pages shows some energy pathways of a salt marsh along the eastern coast of North America.

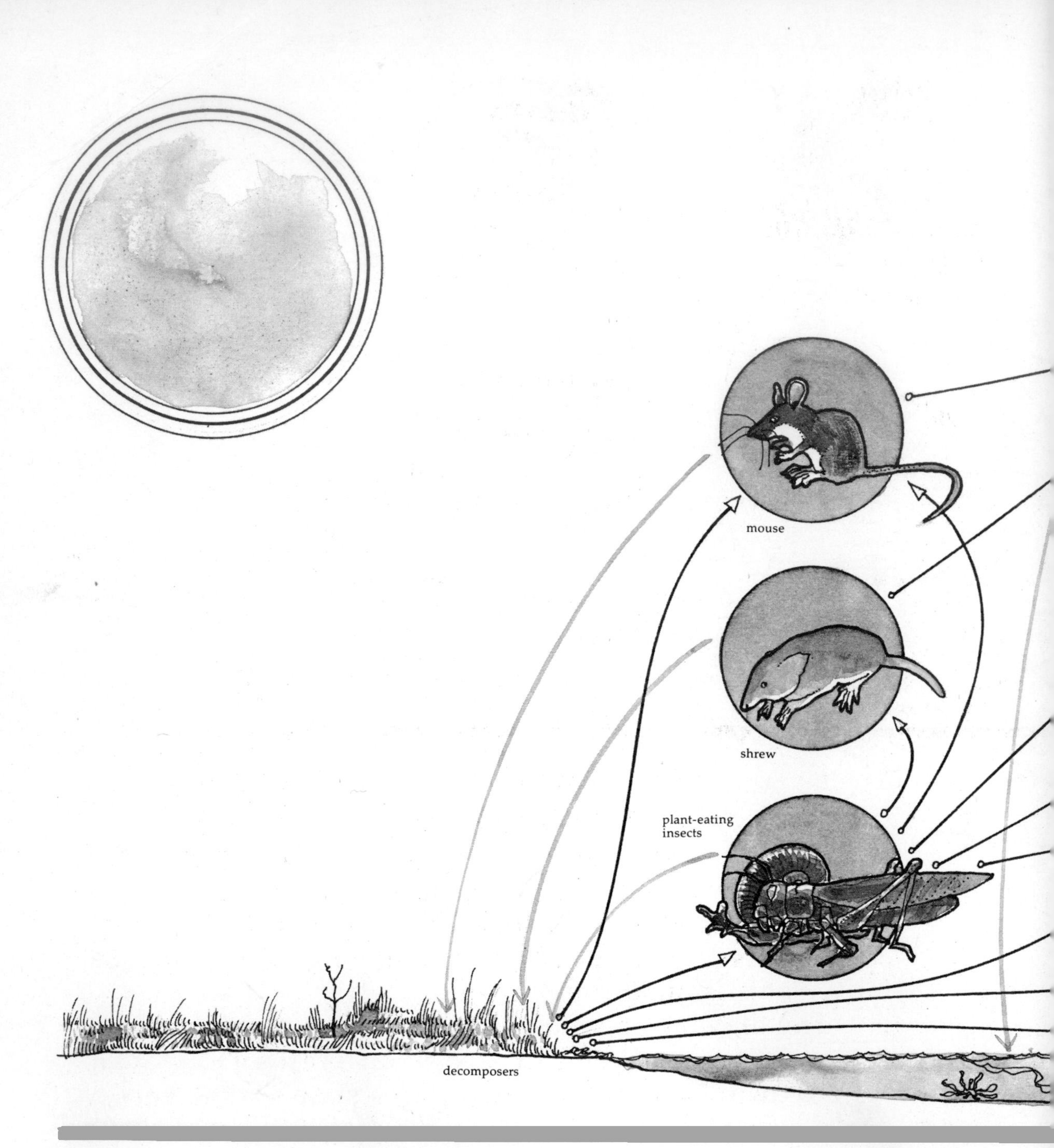

SALT MARSH FOOD WEB

hawk and owl
sandpiper
rat
heron
duck
sparrow
crab and shrimp
fish
decomposers

The flow of energy in nature is of special interest to ecologists—scientists who study the relationships between living things and their surroundings. Ecologists use food webs and chains as models of what they know about energy flow. Ecologists use other models as well, including ecological pyramids.

Like the real pyramids in Egypt, ecological pyramids have broad bases and small tops. They are like food chains turned to point upward. However, they give more information about plant-animal communities than food chains give. Look at this pyramid of numbers, for example.

The base of an ecological pyramid is always made up of the food producers, or green plants. If you were drawing a pyramid of numbers for a field of bluegrass, the base

would represent the total number of grass plants. The next energy level of the pyramid would be the plant eaters. They would be mostly insects but might include some mice. The third level would be the meat eaters, or carnivores. It would be made up of the total numbers of beetles, spiders, ants, and other animals that fed on the plant-eaters. These animals, in turn, might be eaten by a few other carnivores, such as birds and moles, which would make up the next highest level.

Just as food chains have only a few links, ecological pyramids have only a few levels. The energy produced by green plants at the bottom of the pyramid is mostly gone by the fourth or fifth level. There is just enough energy left to support a few carnivores at the top.

Not all ecological pyramids have a pyramid shape. Suppose, for example, you tried to figure out a pyramid of numbers for a maple tree. The bottom or producer level would be easy to count: one tree. From then on, the model would have a normal pyramid shape. Thousands of insects would feed on the maple leaves. Lesser numbers of spiders, insect-eating birds, and other carnivores would be in the higher energy levels of the pyramid.

One big maple tree might produce as much energy as a small field of grass. Numbers alone do not tell how much energy plants and animals contain. So ecologists have found other models more useful than a pyramid of numbers. For example, the weight of a plant or animal is a rough measure of how much energy it contains. Ecologists sometimes figure out pyramids based on the biomass, or total weight, of living things in a community of plants and animals.

Another kind of model used by ecologists is the pyramid of energy. They have found ways to measure the amount of energy produced by plants over the span of a year. Ecologists can also figure out how much of its energy a plant or animal uses up, and how much is left for the animals that eat it. In this way ecologists can draw a pyramid of energy for a plant-animal community.

When people speak of food energy, they use a unit of energy called the calorie. You have probably heard people say, "That dessert tempts me, but it has too many calories." A calorie is the amount of heat energy needed to raise the temperature of one gram of water by one degree centigrade. Ecologists measure energy in calories when they study small organisms or small amounts of energy. But they usually work with large amounts of energy, and use large units called kilocalories. A kilocalorie is a thousand calories of energy. Ecologists try to find out how many kilocalories of energy are produced in one year by plants in a square area measuring one meter on each side (a meter is 39.37 inches). So the energy produced or used by living things is usually measured in kilocalories per square meter per year (kcal/m^2/yr.).

One of the first studies in which energy was measured in an entire plant-animal community took place in Silver Springs, Florida. These beautiful clear-water springs are visited by thousands of tourists each year. The springs are rich with life. Turtles, small fish, and insects feed on algae and freshwater eelgrass. Bigger fish eat the plant eaters, and in turn are fed upon by still bigger fish, such as bass and gar.

The study showed how energy is lost as it flows through food chains. The plants in Silver Springs produced about 20,000 kcal/m^2/yr. of energy. But the few big fish at the top of the energy pyramid received only about 21 kcal/m^2/yr. The fish used most of this energy for themselves. The ecologists figured that only about 6 kcal/m^2/yr. would be available to any animal that fed on the fish.

In Massachusetts, an ecologist named John M. Teal studied the energy flow in a much smaller spring. Root Spring is only about seven feet wide and a few inches deep. Dr. Teal found that algae and other tiny plants in the spring produced about 710 $kcal/m^2/yr$. But more than three times that amount of energy came from outside, mostly from dead leaves that fell into the spring. In fact, most of the insects and other small animals lived among the dead leaves that had settled at the bottom.

Dead leaves fall into ponds, creeks, and rivers, too. So do twigs, flowers, and insects. In this way, streams and ponds receive great amounts of energy from the life along their shores. When leaves and other once-living things decay into little bits and pieces, this material is called detritus. Many water insects in streams and ponds eat detritus.

Most of the animals in Root Spring received their food energy from detritus. Dr. Teal later discovered that many food chains in salt marshes also begin with bits of dead leaves and other detritus. He studied salt marshes along the eastern coast of the United States. There he found that very few of the plants were eaten while they were alive. Most of the marsh animals, including snails, oysters, mussels, and small crabs, fed on bits of dead plants. Then larger animals ate the detritus feeders. Dr. Teal found that about 90 percent of the energy produced by marsh plants flowed through food chains that began with detritus.

The same is true in many other plant-animal communities. In forests, tons of leaves, twigs, and flowers fall to the ground each year. Few animals eat the freshly fallen leaves. However, once the leaves have partly decayed, they are eaten by many animals. Centipedes, snails, and earthworms get energy from dead material on the forest floor. Ecologists have studied the flow of energy in forests in England and the United States. They found that ten times as much energy flows through detritus food chains as through food chains that begin with living plants.

When people think of food chains and pyramids, they usually picture something like grass→cow→human, or acorn→squirrel→owl. We do not usually picture a food chain beginning with bits and pieces of dead leaves. In forests, meadows, marshes, and many other communities, however, most food chains begin in just that way.

As ecologists have learned more about the flow of energy, they have made better models of what happens in plant-animal communities. The drawing on this page shows one such model. A model like this is much more complicated than a simple food chain or pyramid. It shows more accurately what really happens in nature.

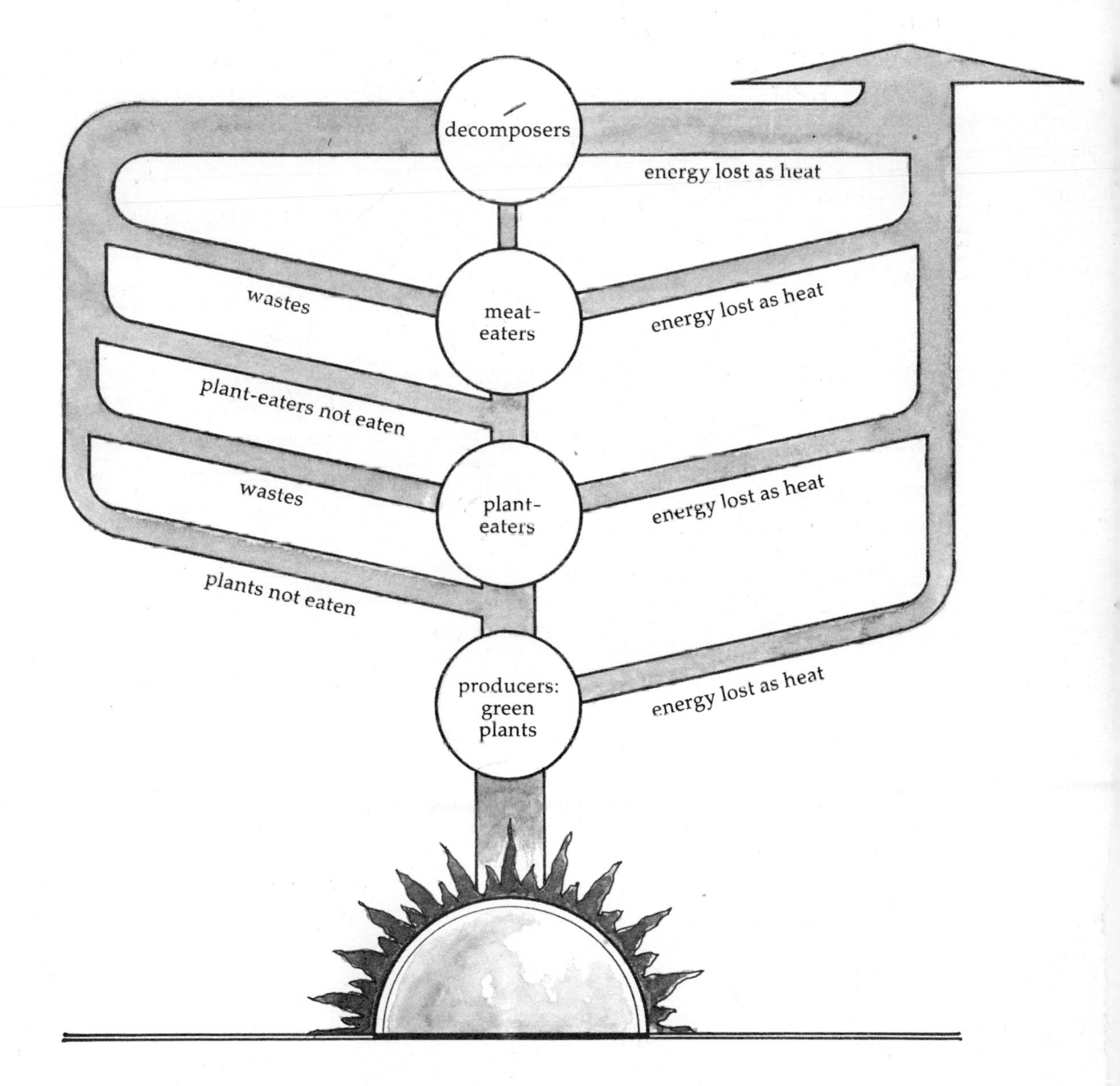

According to the discoveries of ecologists, some plant-animal communities produce much more energy than others. Very little energy is produced in deserts and in the open sea. Desert soils are fertile, but there is not enough rain for much plant growth. In oceans, sunlight reaches only a little way beneath the surface. Below that, there is no sun energy for the drifting algae plants which are the first link in ocean food chains. Also, most ocean waters lack the nutrients needed for normal growth of plants and animals.

Nutrients flow into oceans along the edges of continents, and these waters produce much more food than equal areas of the open sea. Among the richest food-producing areas on earth are estuaries—the places where fresh water from rivers mixes with the salt water of oceans. They are homes for many kinds of fish, and for scallops, clams, oysters, and other seafoods. Some estuaries produce twenty times as much food as an equal area of open sea. Most of the food energy in estuaries grows in grassy salt marshes. Some of these marshes produce more food energy per acre than the best wheat lands on earth.

Other wet places also produce great amounts of food energy per acre. Swamps, marshes, rain forests, shallow lakes, springs, and coral reefs all yield much more energy than deserts or the open sea.

Many people do not know the value of these areas. Tropical rain forests are being cut down and burned. Salt marshes are being polluted and filled in. Other marshes and swamps are often considered "wastelands" and are used as dumps for garbage and trash. Pollution of shallow lakes causes thick growths of algae and other water plants. When the plants die and decay, most of their energy flows to decomposers, instead of to fish that might be used as food by people. This reduces the amount of energy available to humans.

People are worried about the world food supply. There are now four billion people on earth. About two thirds of them are poorly fed. By the year 2000 there may be seven billion people to feed.

Scientists are trying to find ways to produce more food energy for people. One way is to grow varieties of rice, wheat, corn, and other grains that yield big crops. But just getting enough calories from grains does not keep people healthy. A diet of rice or wheat will give you all the calories of food energy you need to keep alive. But you might be crippled or sick because you lacked certain vitamins or minerals. Without enough calcium, for example, your bones and teeth would be weak or poorly formed. Proteins are an especially important kind of nutrient for human health.

Plants contain some protein, but most plants do not contain much high-quality protein—the kind that people need in order to grow and develop properly. Scientists are trying to find ways to get more protein in people's diets. One way is to test many varieties of grains and find some that produce more protein than others.

This was done recently with a grain called sorghum. It is grown in dry soils in poor countries. Only wheat, rice, and corn are more widely used. Scientists tested nine thousand kinds of sorghum from all over the world. They found two kinds that are much richer in protein than the varieties usually planted. If big crops of protein-rich grains such as these can be grown, people will be healthier and

part of the world's food-supply problem will have been solved.

Another way to produce more food is for people to get more of their energy from short food chains. Remember, the shorter the food chain, the less energy lost as heat. Most of our seafood comes from fish, such as cod and tuna, which are carnivores. They are usually the fourth link in food chains. People could get much more protein and food energy if they ate lower on ocean food chains. Scientists are investigating this idea. They are testing ways for harvesting krill, shrimplike animals that live in the Antarctic Ocean. Since krill eat algae, they are only the second link in food chains.

In the United States, each year more than 600 billion tons of grain are fed to animals. The animals produce 84 billion tons of meat, milk, and eggs. In order to produce these protein foods, huge amounts of energy are lost. Energy is lost when grains are fed to livestock and turned into protein. More energy is lost as the milk, meat, and eggs are used by people. Also, many people eat more protein than they actually need.

Poor nations cannot afford to waste much energy on the long food chains needed to raise meat animals. People in India, Asia, and Africa depend mostly on grains for their own food. If the number of people on earth keeps rising, eventually everyone in the world may have to feed lower on food chains. We will eat less meat and depend more on grains, vegetables, and fruits for our food energy.

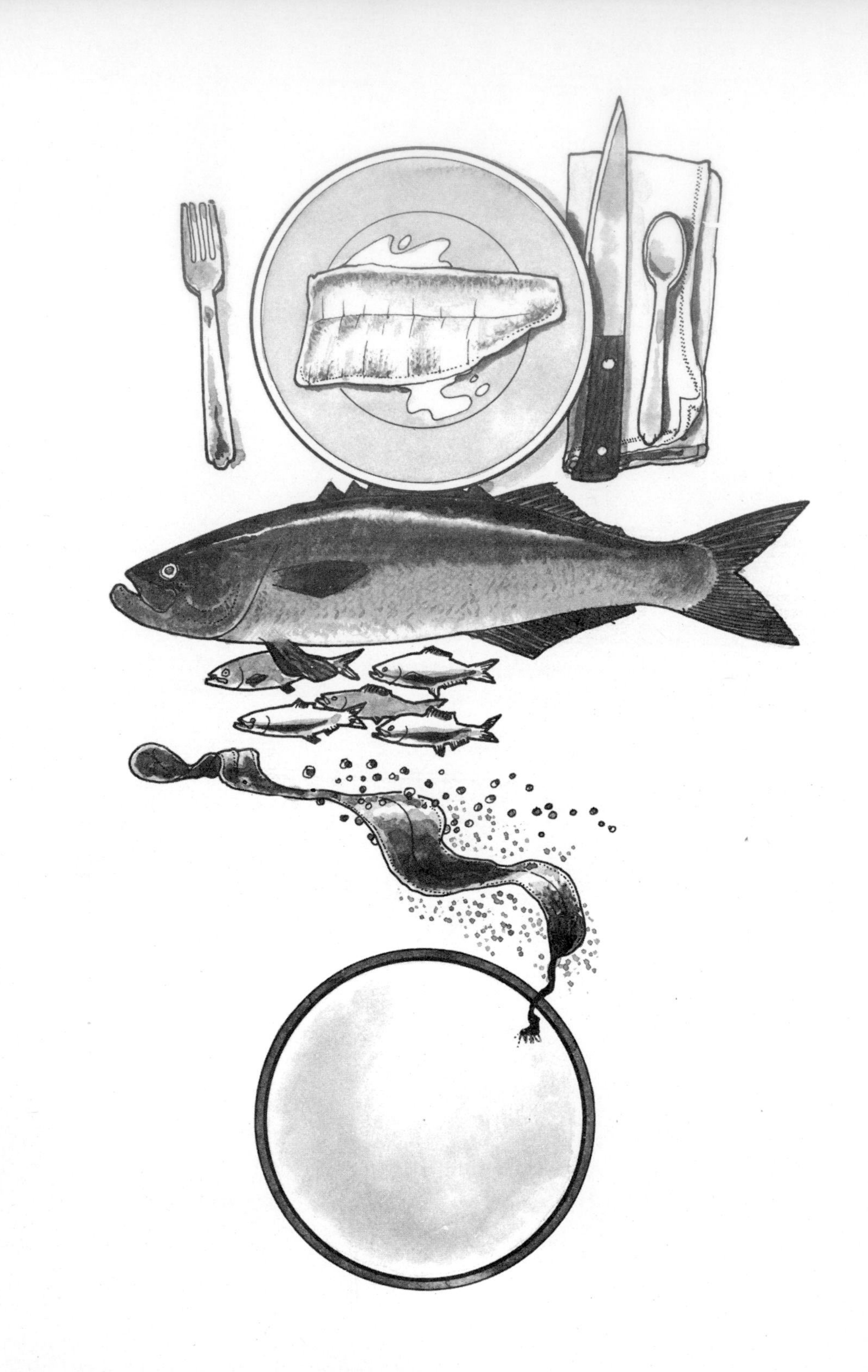

Some people still think that the flow of life's energy happens only in "the wild," far away from kitchens and cafeterias. But humans are just as much a part of energy flow as mice in forests and sharks in the sea. Our lives depend on sunlight and green plants. We are part of many food chains. Our health and survival depend on our understanding of the flow of life's energy on earth.

INDEX

ABOUT THE AUTHOR

Laurence Pringle is the author of twenty books, whose subjects range from dinosaurs and their world to environmental problems today. His enduring interest in nature and ecology has enabled him to supplement scholarly research with firsthand investigation. Mr. Pringle, who has degrees in wildlife conservation from Cornell University and the University of Massachusetts, is a gifted photographer, illustrating many of his books with his own photographs. He shares an acre of land in West Nyack, New York, with katydids, woodchucks, garter snakes, screech owls, and wild flowers.

ABOUT THE ILLUSTRATOR

Jan Adkins is a writer as well as an artist and has produced several handsome books on traditional handicrafts ranging from woodworking to wine-making. His books have received citations from the Brooklyn Museum of Art and have been included in the Children's Book Showcase. In 1972 his *Art and Industry of Sandcastles* was a nominee for a National Book Award.

Mr. Adkins grew up in Ohio but now lives on Cape Cod, where he writes and draws and sails and walks, and fools around.